AF095676

THE POETRY OF OXYGEN

The Poetry of Oxygen

Walter the Educator

SKB

Silent King Books a WhichHead Imprint

Copyright © 2023 by Walter the Educator

All rights reserved. No part of this book may be reproduced in any manner whatsoever without written permission except in the case of brief quotations embodied in critical articles and reviews.

First Printing, 2023

Disclaimer
This book is a literary work; poems are not about specific persons, locations, situations, and/or circumstances. This book is for entertainment and informational purposes only. The author and publisher offer this information without warranties expressed or implied. No matter the grounds, neither the author nor the publisher will be accountable for any losses, injuries, or other damages caused by the reader's use of this book. The use of this book acknowledges an understanding and acceptance of this disclaimer.

dedicated to all the chemistry lovers like myself

CONTENTS

Dedication v

Why I Created This Book? 1

One - Gives Life 2

Two - Element Divine 4

Three - Human Grace 6

Four - Gentle Embrace 8

Five - Our Very Breath 10

Six - Abundant Might 12

Seven - Grand Symphony 14

Eight - Till Death 15

Nine - The Tapestry Of Life 17

Ten - Every Breath Spent 19

Eleven - Life Forevermore 21

Twelve - Life's Sweet Breath 23

Thirteen - Rhythm Divine	. . . .	25
Fourteen - Making Us Whole	. . . .	27
Fifteen - Symbol Of Life	. . . .	29
Sixteen - Oxygen, Oh Oxygen	. . . .	31
Seventeen - Intricate Design	. . . .	32
Eighteen - Fill Your Lungs	. . . .	34
Nineteen - Earth Sings	. . . .	36
Twenty - Invisible Gas	. . . .	38
Twenty-One - Every Heartbeat	. . . .	39
Twenty-Two - Element So Pure	. . . .	41
Twenty-Three - Truly Derives	. . . .	43
Twenty-Four - Veins It Roams	. . . .	45
Twenty-Five - Element Of Life	. . . .	47
Twenty-Six - Waltz In The Air	. . . .	48
Twenty-Seven - Precious Element	. . . .	50
Twenty-Eight - Silent Force	. . . .	51
Twenty-Nine - Giver Of Life	. . . .	53
Thirty - Fills Us With Love	. . . .	54
Thirty-One - Lifeline We Hold	. . . .	56
Thirty-Two - Through And Through	. . . . . .	57

Thirty-Three - Pure And Divine	. . . .	59
Thirty-Four - Interconnectedness On Earth	. . . .	61
Thirty-Five - Body And Soul	. . . .	63
Thirty-Six - Chemistry's Art	. . . .	64
Thirty-Seven - Power To Last	. . . .	66
Thirty-Eight - Delicate Thread	. . . .	67
Thirty-Nine - Life-giving Gas	. . . .	69
About The Author	. . . .	70

WHY I CREATED THIS BOOK?

Creating a poetry book about the chemistry element of Oxygen is a unique and captivating way to explore the scientific concept through artistic expression. Oxygen, as an essential element for life, offers a plethora of metaphors and symbolism that can be beautifully woven into poetry. By delving into the properties, functions, and significance of Oxygen, I can create a collection that explores themes of breath, vitality, transformation, and interconnectedness. This interdisciplinary approach not only fosters a deeper understanding of science but also demonstrates the beauty and interconnectedness of different fields of knowledge.

ONE

GIVES LIFE

In the realm of elements, a celestial star,
A giver of life, both near and far,
Oxygen, the breath of existence,
A divine essence, a vital persistence.

A diatomic dance, two atoms entwined,
Creating a bond, a union defined,
With electrons shared, they form a pair,
Oxygen's presence fills the air.

From the depths of the ocean to skies up high,
Oxygen's abundance, never to deny,
It fuels the flames with a fervent might,
A catalyst for fire, a source of light.

In every living being, from creature to plant,
Oxygen sustains, a gift to enchant,
It fuels the cells, ignites the fire,
A source of energy, a life's desire.

With every breath, a sacred exchange,
Oxygen rejuvenates, it does not estrange,
It nourishes the body, invigorates the soul,
A life-giving force, making us whole.

In the chemistry of life, it plays its part,
From photosynthesis to the beating heart,
Oxygen, the element, so pure and true,
A symbol of life, a gift to pursue.

So let us celebrate this element divine,
Oxygen, the essence, that makes us shine,
With gratitude and awe, we raise our voice,
For the element that gives life, we rejoice.

TWO

ELEMENT DIVINE

In the realm of elements, one shines bright,
A vital force that fuels the flames so white.
Oxygen, divine essence of the air,
A breath of life, a gift beyond compare.

With every gasp, our lungs embrace its touch,
Reviving cells, invigorating much.
From the highest peaks to the deepest sea,
Oxygen, you nourish us with glee.

In fiery dance, you ignite the blaze,
A catalyst for life in countless ways.
From stars to candles, you sustain the fire,
A force of nature that we all admire.

You bind with carbon, creating life's chain,
A symphony of molecules, a sweet refrain.
In water's embrace, you lend your hand,
A vital component of life's grand band.

Oh, Oxygen, we owe you our gratitude,
For the air we breathe, the life you include.
You're the essence that keeps our souls alive,
A symbol of hope, as we strive to thrive.
  So let us celebrate this element divine,
A gift of nature, a treasure so fine.
Oxygen, we cherish your presence here,
With every breath, our gratitude is clear.

THREE

HUMAN GRACE

In the dance of molecules, a star is born,
A vital force that fuels life's every dawn.
Oxygen, the breath of air we all seek,
An elemental symphony, humble and meek.

From the depths of the oceans to the skies above,
You bless us with the gift of life, pure love.
Through your embrace, fires are ignited,
Burning bright, your touch never slighted.

In every gasp, you fill our lungs with might,
A symphony of atoms, a celestial light.
With each exhale, you carry our dreams,
A lifeline of hope, woven in oxygen's streams.

You sustain the body, a constant friend,
A healer and protector, until the end.
When darkness looms and shadows creep,
Oxygen, you are the promise we keep.

In every heartbeat, you whisper a song,
A rhythm that guides us, steady and strong.
You bind us together, a thread of life's tapestry,
Oxygen, our eternal source of harmony.
So let us cherish your presence, so divine,
For without you, life's symphony would decline.
In every breath, we taste your sweet embrace,
Oxygen, the essence of our human grace.

FOUR

GENTLE EMBRACE

In the realm of elements, majestic and pure,
There exists a force that makes life endure.
Oxygen, the breath of life it brings,
A vital force that sustains all living things.

From the depths of the ocean to the blue skies above,
Oxygen whispers its presence with grace and love.
It binds with carbon, creating energy's flame,
Fueling the cycle of life, forever the same.

With every inhale, we welcome its might,
A symphony of atoms, dancing in the light.
Through our lungs, it courses, giving us strength,
Revitalizing our bodies, a gift of immense.

Oxygen, oh noble element divine,
Without you, life's symphony would never align.

We owe our existence to your gentle embrace,
Forever grateful for your life-giving grace.

FIVE

OUR VERY BREATH

In the realm of atoms, it reigns supreme,
A life-giving force, a celestial dream.
Oxygen, the element of breath and fire,
Fueling the flames of our heart's desire.

From the depths of space to the earthly skies,
It whispers its secrets, a gift in disguise.
Without its touch, life's spark would cease,
A symphony of existence, a delicate peace.

It weaves through our veins, a crimson dance,
Nourishing every cell with its gentle advance.
In every inhale, a moment to embrace,
The essence of life, its eternal grace.

It binds with carbon, a cosmic embrace,
Creating the tapestry of life's fertile space.
In every molecule, a story untold,
Of Oxygen's journey, a tale of old.

From the first breath of life to the final sigh,
It accompanies us, never asking why.
Oh, Oxygen, we owe you our very breath,
In your presence, we find life's blessed depth.

SIX

ABUNDANT MIGHT

In the realm of elements, a breath of life,
Oxygen, the giver, banisher of strife.
A silent hero, dwelling in the air,
Without you, existence would be unfair.
Through countless ages, you've remained the same,
A faithful companion in nature's game.
In every molecule, a vital role you play,
From lofty skies to depths of ocean's sway.
You join with carbon, hydrogen, and more,
Creating life's tapestry, a boundless score.
From the tallest trees to the tiniest cells,
Oxygen, your presence, in every tale it tells.
With each inhale, you fill our lungs with grace,
Revitalizing every part of our earthly space.
You fuel the fire of life's eternal dance,
A catalyst for growth, a chance to enhance.

Oxygen, the essence of our very breath,
A precious gift that conquers life's cold death.
We owe our thanks to your abundant might,
For without you, darkness would shroud our light.

SEVEN

GRAND SYMPHONY

In nature's realm, a gift bestowed,
A breath of life, a vital code.
Oxygen, the cosmic fire,
Ignites the flames of heart's desire.

Within our lungs, it finds its way,
To fuel the cells that dance and sway.
A catalyst for life's grand scheme,
A vibrant thread in nature's dream.

It binds with atoms, strong and true,
Creating bonds, a cosmic glue.
With carbon, hydrogen, and more,
It weaves the tapestry we adore.

From fiery stars to ocean's crest,
Oxygen's touch, we are blessed.
A force of nature, wild and free,
Revealing life's grand symphony.

EIGHT

TILL DEATH

In the vast expanse of cosmic night,
There shines a celestial light,
An element of life's grand chore,
Oxygen, the essence we adore.

From the stars it first did form,
A gift to Earth, in a celestial storm,
With every breath, we take it in,
A lifeline of hope, a virtue within.

In every molecule, a bond it weaves,
A giver of life, our souls it frees,
With fiery fervor, it fuels the flame,
Oxygen, the element that bears no shame.

From the depths of oceans to mountain peaks,
It nurtures life, it's what we seek,
In every forest, in every glen,
Oxygen breathes, the savior of men.

So let us cherish this element divine,
For in its presence, we shall shine,
A tribute to the giver of breath,
Oxygen, the elixir of life, till death.

NINE

THE TAPESTRY OF LIFE

Oxygen, the giver of fire,
A catalyst for life's desire,
Its electrons dance in a cosmic choir,
And its bond with life will never tire.

It fuels the flames of passion and love,
And powers the engines that move above,
It's in every breath we take, like a dove,
And in every step we make, like a glove.

Oxygen, the secret of the stars,
The breath of life that's never far,
It's the bridge between worlds, near and far,
And it's the light that shines like a jar.

It's the element that sustains our breath,
And the force that drives us to our depth,

It's the key to life, like a precious wreath,
And the essence of existence, like a stealth.
    Oxygen, the tapestry of life,
The thread that connects us like a knife,
It's the bond that weaves us through strife,
And the harmony that makes us thrive.
    It's the element that joins us all,
And the force that makes us stand tall,
It's the essence of being, like a call,
And the music of nature's grand ball.

TEN

EVERY BREATH SPENT

In every breath, a whispered sigh,
A vital force that keeps us alive.
Oxygen, the element of life,
A sustainer of all, in its grandest stride.

With every beat, the heart does pound,
Oxygen, the fuel that makes it resound.
From the air we breathe, it finds its way,
To every cell, it brings a brighter day.

In the forest's embrace, it weaves its spell,
Through leaves and boughs, it dances so well.
Enriching the Earth with its gentle touch,
Oxygen, the giver, it means so much.

From the depths of the sea to the soaring skies,
Oxygen's presence, a constant surprise.
It binds with other elements, a cosmic embrace,
Creating the tapestry of life, in every place.

So let us marvel at this wondrous element,
That keeps us alive, with every breath spent.
Oxygen, the giver of life's sacred flame,
Forever bound to us, in its eternal claim.

ELEVEN

LIFE FOREVERMORE

In the realm of atoms, a celestial dance,
Resides a noble element, a cosmic romance.
Oxygen, the life-giver, with every breath we take,
Sustains our existence, a bond no one can break.

From the depths of the oceans to the highest peak,
It intertwines with nature, gentle and meek.
Inhaled by the lungs, it fuels the beating heart,
A symphony of life, a masterpiece of art.

With hydrogen it combines, a union so pure,
Creating water, a source of life secure.
In the fiery stars, it fuels their brilliant light,
A cosmic embrace, a celestial delight.

Oxygen, the giver of life's vital force,
From the smallest organisms to the mightiest horse.
It weaves through the tapestry of Earth's design,
A precious element, forever intertwined.

So let us cherish this gift from the skies above,
For without oxygen, there would be no love.
In every breath we take, its presence we adore,
Oxygen, the essence of life forevermore.

TWELVE

LIFE'S SWEET BREATH

In the realm of elements, Oxygen stands,
A giver of life, in nature's hands.
It dances with Carbon, a duet so grand,
Building the fabric of life's endless strand.

Deep in the oceans, where creatures reside,
Oxygen bubbles, a lifeline supplied.
From gills to lungs, it breathes in the air,
Sustaining existence, with tender care.

In forests, it whispers among the trees,
A vital ingredient, for all to seize.
Through photosynthesis, it fuels the Earth,
Green leaves embrace it, a divine rebirth.

Oxygen, the sustainer of flame,
Igniting the fire, a powerful game.
From candles to stars, its magic unveiled,
A spark of creation, forever hailed.

High in the sky, it forms a protective shield,
Shielding us from harm, a force to wield.
Ozone, its guardian, a shield against strife,
Preserving the balance of our fragile life.

In every breath we take, Oxygen we find,
A precious elixir, intertwined.
From birth to death, it lingers within,
A constant companion, through thick and thin.

So let us cherish this element of grace,
Embrace its essence, in every place.
For Oxygen, the giver of life's sweet breath,
Connects us all, in this dance till death.

THIRTEEN

RHYTHM DIVINE

In the realm of elements, a celestial gem,
Oxygen, the lifeline, the force that fuels the flame.
With every breath, it dances through our veins,
Nourishing the essence, from whence life came.

A giver of energy, it powers our might,
Oxygen, the elixir, that keeps us alive.
From the depths of the sea to the heights of the sky,
It whispers its secret, a breath we can't deny.

With open arms, it welcomes all with grace,
Oxygen, the bridge, between worlds we embrace.
Through its invisible threads, connections are made,
Binding us together, in a vast cosmic parade.

It breathes life into dreams, igniting the fire,
Oxygen, the essence, of existence's desire.
From birth to the last exhale, it remains,
A constant companion, amidst joys and pains.

Oh, Oxygen, the tapestry of life's grand design,
We owe our every beat, to your rhythm divine.
In every molecule, a story unfolds,
A bond that connects us, as the tale of life unfolds.

FOURTEEN

MAKING US WHOLE

Oxygen, the element of air,
A vital link in life's grand affair.
In every breath, we take it in,
A silent force that keeps us within.

It binds us to the earth we tread,
And to the skies, we look ahead.
It's the bridge between worlds, we see,
The light that shines, forever free.

Oxygen, the thread that connects us,
The essence of existence, it fusses.
It flows through our veins and arteries,
And gives us the strength for life's vagaries.

It's the force that drives us on,
The key to life, from dusk to dawn.
It sustains our breath, and our soul,
And weaves us through strife, making us whole.

Oxygen, the giver of life's sacred flame,
Forever bound to us, in nature's game.
It's the element that sustains our every breath,
And the bond that weaves us, even in death.

FIFTEEN

SYMBOL OF LIFE

Oxygen, the bridge between worlds,
Invisible, yet ever-present,
A catalyst of life, untold,
In every breath, a gift sent.

It dances in the atmosphere,
A vital part of Earth's embrace,
From mountain peaks to oceans clear,
Oxygen connects every place.

Through bonds that hold, it weaves a thread,
Linking atoms in a grand design,
Fueling fires with flames widespread,
Oxygen, the force that defines.

It whispers secrets to the stars,
Guiding comets on their cosmic flight,
With each exhale, it heals the scars,
Oxygen, the breath of pure light.

So let us cherish this precious gas,
For it sustains our every breath,
Oxygen, the life-force that will last,
A symbol of life, conquering death.

SIXTEEN

OXYGEN, OH OXYGEN

Oxygen, oh Oxygen,
You're the breath of life, my friend.
You sustain us with every inhale,
And without you, we would surely fail.
But you're more than just a gas,
You connect everything in the past.
From the trees that sway in the breeze,
To the stars that twinkle with ease.
You're a part of every molecule,
A force that keeps everything cool.
You guide the winds and shape the land,
A vital element, forever in demand.
Oxygen, oh Oxygen,
You're the force that keeps us goin'.
Thank you for all that you do,
We couldn't survive without you.

SEVENTEEN

INTRICATE DESIGN

In the realm of atoms, a wondrous tale unfurls,
Of a humble element, nature's breath, it swirls.
Oxygen, the giver of life, the vital link,
Connecting all beings, in perfect sync.

From the depths of the ocean to the highest peak,
Oxygen whispers, as it flows, so sleek.
It dances through the air, unseen and pure,
A force that sustains, a miracle to endure.

In every breath we take, it fills our lungs,
A symphony of life, where every note is sung.
From the smallest creature to the mightiest tree,
Oxygen weaves its magic, setting us free.

It fuels the raging fires and lights up the night,
With every spark, it ignites our inner light.
In chemistry's embrace, it forms bonds so strong,
Uniting elements, where harmony belongs.

Oxygen, the catalyst of energy's flame,
A molecule of wonder, we cannot name.
It binds and combines with atoms divine,
Creating life's tapestry, an intricate design.

So let us cherish this element so dear,
For without oxygen, life would disappear.
In every breath we take, let gratitude arise,
For oxygen's embrace, the gift that never dies.

EIGHTEEN

FILL YOUR LUNGS

In every breath we take, a silent dance,
Oxygen, the element that gives life a chance.
From the depths of the ocean to the sky above,
It connects us all, with an invisible love.

In every leaf that rustles, every bird that soars,
Oxygen whispers, as life's symphony roars.
It fuels our bodies, ignites the flame,
A catalyst for energy, in this cosmic game.

From the first cry of a newborn to the last sigh,
Oxygen sustains us, as time passes by.
In each beat of our hearts, a rhythm so true,
Oxygen weaves its magic, binding me to you.

It circulates through rivers, it flows through trees,
Oxygen connects, in every gentle breeze.
It's the spark of inspiration, the fire in our soul,
With Oxygen, we're part of a greater whole.

So breathe it in deeply, let it fill your lungs,
Oxygen, the element that ties us to the young.
In its embrace, we find strength and grace,
A reminder that life's beauty is in every breath we chase.

NINETEEN

EARTH SINGS

In every breath, a sacred bond,
Oxygen, the life's beyond.
The giver of life, a vital force,
Connecting all, a binding source.

Within our lungs, you find your way,
Reviving cells, with each new day.
From atmosphere to bloodstream flow,
A rhythm of life, you help bestow.

From mountain peaks to ocean deeps,
You touch all life, where'er it creeps.
You fuel the fire in our hearts,
And spark the flame when life restarts.

In every leaf, you play a part,
A gift of life, a work of art.
Through photosynthesis, you weave,
The tapestry of plants, you conceive.

From greenest forests to garden's bloom,
You paint the world with vibrant plume.
The breath of life, the pulse of Earth,
Invisible yet of infinite worth.

Oxygen, the element divine,
With every beat, your presence shines.
A bridge between all living things,
In you, the harmony of Earth sings.

TWENTY

INVISIBLE GAS

Invisible gas that we breathe,
Oxygen sustains all living beings,
From the tiniest microbe to the tallest tree,
It fuels the cycle of life, keeping us all free.
 It rusts the iron and burns the flame,
It makes up water and helps plants gain fame,
Its electrons are shared, forming bonds so strong,
Without it, life would simply be gone.
 It powers our cells, helps us to think,
It's the spark in our body that helps us blink,
So let us cherish this simple gas,
For without it, our world would come to pass.

TWENTY-ONE

EVERY HEARTBEAT

In every breath, a hidden grace,
The element that fuels life's pace.
Oxygen, the giver of fire,
Ignites the soul with burning desire.

From the azure sky to the deepest sea,
Oxygen connects all living beings.
In every leaf, a vibrant hue,
A testament to the air we breathe through.

It binds us together, this element divine,
From the tallest tree to the crawling vine.
Through photosynthesis, it lends a hand,
Creating energy in nature's grand plan.

With carbon, hydrogen, and nitrogen,
Oxygen dances, a chemical origin.
It forms the bonds that make life thrive,
A symphony of atoms, keeping us alive.

In every heartbeat, a rhythmic beat,
Oxygen pulses in our veins, so sweet.
It nourishes our cells, a vital force,
Sustaining life's journey, a boundless course.

So let us cherish this element rare,
For without oxygen, life would be bare.
In every breath, let gratitude rise,
For oxygen's gift, a world that never dies.

TWENTY-TWO

ELEMENT SO PURE

Oxygen, the life-giving gas,
Invisible but vital, it does pass
Through our lungs and to each cell,
A necessity for us to live well.

But beyond our human need,
Oxygen connects all living beings indeed.
From green plants to the smallest microbe,
It powers the cycle of life, don't you know?

Oxygen, a giver of light,
Is present in stars, burning bright.
It bonds with hydrogen to form water,
A force of nature, a life supporter.

So let us cherish this element so pure,
For without it, life could not endure.
In every breath we take, let us give thanks,

For oxygen sustains us, from the smallest plankton to the biggest banks.

TWENTY-THREE

TRULY DERIVES

In the realm of atoms, a wondrous tale unfolds,
Of a vital element, a story yet untold.
Oxygen, the breath of life, so pure and divine,
Connecting all beings in a mystical design.

From the depths of the ocean to the heights of the sky,
Oxygen's presence, oh so crucial, cannot deny.
It fuels the fiery dance of the stars above,
And whispers secrets of the universe with love.

With every inhale, we invite it into our chest,
Oxygen, the giver of life, we are truly blessed.
It binds with carbon, forming life's sacred bond,
In every living cell, its essence is fond.

It weaves through the veins, a lifeline so true,
Oxygen, the sustainer, our journey it continues.

In every heartbeat, it pulses with grace,
Nourishing our bodies, leaving no trace.
 Oxygen, the alchemist, transforming the air,
Breathing life into existence, with tender care.
As we gaze at the sky, filled with stars so bright,
We marvel at oxygen's role, in this cosmic delight.
 In every breath we take, let gratitude reside,
For oxygen, the element that keeps us alive.
A symphony of atoms, in harmony it thrives,
Oxygen, the conductor, where life truly derives.

TWENTY-FOUR

VEINS IT ROAMS

In nature's grand plan, a miracle resides,
A vital element, where life abides.
Oxygen, the breath of life it brings,
Enabling creatures to flutter on wings.

In each breath we take, a gift so pure,
Oxygen dances, a force to endure.
It fuels the fire, ignites the spark,
A symphony of life, a celestial arc.

From depths of oceans to mountain's peak,
Oxygen flows, in every molecule we seek.
It binds with carbon, creating the breath,
The essence of life, the cycle of death.

In verdant forests, it whispers through trees,
A gentle caress, carried on the breeze.
It nourishes the Earth, with each exhale,
A symbiotic dance, an eternal tale.

In lungs of creatures, it finds a home,
A vital elixir, through veins it roams.
From birth to twilight, every living soul,
Depends on oxygen, to keep them whole.

So let us cherish this element divine,
A gift from nature, so precious, so fine.
For oxygen is the thread that weaves,
The tapestry of life, where hope achieves.

TWENTY-FIVE

ELEMENT OF LIFE

Invisible and essential,
Oxygen is what we breathe,
A gas so vital,
Without which we'd seethe.
It's present in the air,
And in water too,
It fuels the flames,
And makes life anew.
It's found in the stars,
And in our very cells,
A bond so strong,
It's hard to dispel.
Oxygen, oh oxygen,
You're the element of life,
You sustain us all,
And keep our world rife.

TWENTY-SIX

WALTZ IN THE AIR

In the realm of elements, Oxygen stands tall,
A silent hero, the giver of life to all.
With every breath, we inhale its embrace,
A vital force, sustaining our human race.

Invisible and weightless, yet a potent force,
Oxygen dances through nature's course.
It fuels the flames of life with its fiery breath,
Igniting the spark of existence, defeating death.

In the depths of the ocean, where life takes form,
Oxygen bubbles rise, a symphony in the storm.
From the towering trees to the smallest blade of grass,
Oxygen whispers in every living thing that passes.

It binds with the carbon, creating life's dance,
A delicate balance, a cosmic romance.
With hydrogen, it forms a waltz in the air,
Water, the elixir of life, beyond compare.

In the quiet night sky, stars twinkle and gleam,
Their light travels through space, a celestial dream.
But without Oxygen, their brilliance would fade,
A reminder that life's breath cannot be betrayed.

So let us cherish this element, so pure and true,
For without Oxygen, there would be no me or you.
In every breath we take, its presence we feel,
A testament to the power of life's eternal wheel.

TWENTY-SEVEN

PRECIOUS ELEMENT

 Oxygen, the breath of life,
A vital element for all to thrive,
In every breath, we take in its essence,
And exhale it out with a sense of presence.

 It's found in the air we breathe,
And in the oceans and majestic seas,
It's a part of every living thing,
From the smallest bug to the mighty king.

 Oxygen fuels the fire of our souls,
And helps us to reach our goals,
It's a chemical bond that we all share,
A reminder that we're all connected, everywhere.

 So let us cherish this precious element,
And the vital role it plays,
For without it, life would cease to exist,
And our world would be a barren place.

TWENTY-EIGHT

SILENT FORCE

In the realm of chemistry, a noble gas so true,
A vital element that brings life anew.
Oxygen, the breath of all that is alive,
From the deepest oceans to the skies up high.

Within our lungs, it forms a sacred bond,
Fueling the fire that keeps us strong.
From every breath, our bodies do receive,
The gift of life, so precious, we believe.

In nature's dance, it weaves a seamless thread,
A silent force that keeps all creatures fed.
From mighty forests to the smallest bloom,
Oxygen's touch, a life-giving perfume.

It fuels the flames that light the darkest night,
A catalyst for fire, burning ever bright.
From flickering candles to the sun's golden rays,
Oxygen ignites the world in wondrous ways.

So let us cherish this element divine,
For without its presence, we cannot shine.
In every breath we take, let's be aware,
Of oxygen's gift, beyond compare.

TWENTY-NINE

GIVER OF LIFE

Oxygen, the giver of life,
Found in the air, water, and stars so bright.
It's a part of every breath we take,
A necessary element we cannot forsake.
But beyond its role in sustaining life,
Oxygen has another side.
It fuels the fire of our souls,
And gives us strength to reach our goals.
In every molecule of our being,
Oxygen is the key to our seeing.
It's the spark that ignites the flame,
And keeps us going in life's game.
So let us celebrate this element divine,
For it has the power to make us shine.
Oxygen, the fuel that keeps us strong,
And the catalyst for life's endless song.

THIRTY

FILLS US WITH LOVE

In the realm of life, oxygen does reside,
A vital element, our very guide.
Within us it dwells, a spark so pure,
Igniting the fire that forever endures.

With each breath we take, it fuels the flame,
Oxygen, the catalyst, our life's mainframe.
From the highest peak to the deepest sea,
Oxygen sustains all living, wild and free.

It dances through the bloodstream, vibrant and bright,
Revitalizing cells, like stars in the night.
A force of nature, invisible yet profound,
Oxygen, the life force that knows no bound.

It whispers in the wind, a gentle breeze,
Nourishing the earth, the flowers and trees.

It roars in the storm, a thunderous sound,
Oxygen, the essence that keeps our world round.
    In every gasp of air, we find our strength,
Oxygen, the elixir that spans the length,
Of time and space, the very fabric of our being,
Oxygen, the element that keeps us from fleeing.
    So let us cherish this gift from above,
Oxygen, the element that fills us with love.
For without its presence, life would cease,
Oxygen, the source of our eternal peace.

THIRTY-ONE

LIFELINE WE HOLD

In the realm of atoms, a beauty profound,
A messenger of breath, where life is found,
Oxygen, the element, with power untold,
Fueling the fires, as ancient stories unfold.

In every breath we take, it whispers its name,
Sustaining our being, a life-giving flame,
From the depths of the ocean, to the sky up above,
Oxygen dances, with grace and with love.

A catalyst for dreams, it ignites the fire,
Fueling the passions, that make us aspire,
With every beat of the heart, it fuels the soul,
Oxygen, the element, that makes us whole.

So let us cherish this gift, so pure and true,
For without its presence, what would we do?
Oxygen, the giver, the lifeline we hold,
In its embrace, our stories unfold.

THIRTY-TWO

THROUGH AND THROUGH

In the realm of life's fiery dance,
Where passions burn and dreams enhance,
There lies a force, unseen and true,
A breath of life, it's Oxygen, through and through.

From the depths of time to the present day,
Oxygen has paved the way,
Nourishing flames that ignite our souls,
Fueling the fire that forever rolls.

With each inhale, we take it in,
A sacred bond, we cannot rescind,
It fills our lungs, and fuels our fire,
A constant presence, never to tire.

In every beat of our hearts, it plays a part,
Oxygen, the rhythm of life, an eternal art,

It binds us all, in a cosmic embrace,
Sustaining existence, with its gentle grace.
    So let us cherish this element divine,
For without Oxygen, life cannot shine,
In every breath, let us give thanks,
For the gift of life, Oxygen forever ranks.

THIRTY-THREE

PURE AND DIVINE

In the realm of atoms, a wondrous tale unfolds,
Of an element vital, a story yet untold.
Oxygen, the breath of life, the giver of flame,
A force that fuels our passions, igniting our name.

In every breath we take, its presence we find,
A bond with the universe, forever intertwined.
From the depths of the oceans to the skies up above,
Oxygen whispers secrets of life, of hope, and love.

It dances with carbon, forming bonds strong and true,
Creating the building blocks of all that we knew.
From the vibrant foliage to the creatures that roam,
Oxygen weaves the tapestry of nature's grand poem.

Its power is boundless, its influence profound,
A catalyst for change, in every corner found.
From the fire's fierce embrace to the stars' gentle

glow,
Oxygen breathes life, wherever it may go.
    So let us cherish this element, so pure and divine,
For without its touch, life's symphony would decline.
In every beat of our hearts, in every breath we take,
Oxygen sustains us, our souls forever awake.

THIRTY-FOUR

INTERCONNECTEDNESS ON EARTH

In the realm of atoms and compounds, behold,
A celestial element, captivating and bold.
Oxygen, the life-giving force of the air,
A presence so vital, beyond compare.

With every breath, we inhale its grace,
Fueling our passions, igniting the chase.
It courses through our veins, a sacred fire,
Empowering our dreams to aim higher.

From the depths of the ocean to the mountain's height,
Oxygen sustains the spark of life's delight.
It dances with carbon, creating endless forms,
Connecting us all, weathering the storms.

In every living creature, it finds a home,
Nourishing our bodies, wherever we may roam.

Oxygen, the essence that keeps our world round,
A symphony of molecules, beautifully profound.
   So let us cherish this gift from above,
For without oxygen, what would life be, my love?
In its embrace, we find strength and rebirth,
A reminder of our interconnectedness on Earth.

THIRTY-FIVE

BODY AND SOUL

In the realm of life, a spark ignites,
A breath of air, a soul takes flight.
Oxygen, the giver of life's breath,
Fueling our passions, conquering death.
From the depths of oceans to the skies above,
Oxygen sustains, it's the essence of love.
In every heartbeat, it is there,
A vital force, beyond compare.
In every flame that dances bright,
Oxygen fuels the fire's light.
It binds with carbon, ignites the heat,
Creating energy, so complete.
Oh oxygen, the element divine,
A catalyst for life's design.
With every breath, we are made whole,
A symphony of atoms, body and soul.

THIRTY-SIX

CHEMISTRY'S ART

In the realm of chemistry's art,
Where elements dance and play their part,
There lies a force that fuels the fire,
An essence that ignites desire.

Oxygen, the breath of life,
Gifted to us from starry heights,
It fills our lungs with every breath,
And whispers secrets of life and death.

It binds with carbon, strong and true,
Creating compounds, old and new,
From water's gentle, flowing streams,
To fiery sparks in lovers' dreams.

It weaves the tapestry of nature's scene,
From lush green forests to fields of green,
In every breath, a sacred dance,
Oxygen's embrace, a cosmic romance.

So let us cherish this vital gas,
For without it, life would surely pass,
In every beat of our hearts, we find,
Oxygen's love, forever intertwined.

THIRTY-SEVEN

POWER TO LAST

    Oxygen, the breath of life,
Invisible force that fuels our strife,
It ignites the fire of our passion,
And drives us forward with fierce attraction.
    It's the spark that lights the flame,
A catalyst that we cannot tame,
It's the force that binds us together,
And gives us life, now and forever.
    Oxygen, the element of the air,
It's the breath that we all share,
In every moment, in every place,
It's the force that sustains the human race.
    So let us honor this vital gas,
And never forget its power to last,
For without it, we could not survive,
Oxygen, the force that makes us thrive.

THIRTY-EIGHT

DELICATE THREAD

In the realm of chemistry's embrace,
There lies a force that ignites our grace.
Oxygen, the breath of life, so pure,
A catalyst that makes our hearts endure.

With every inhale, it fuels our fire,
Within our lungs, it never tires.
It binds with carbon, forming energy's dance,
Creating life's rhythm, a divine romance.

Oxygen, the element of passion,
Through our veins, it flows in fashion.
It fuels the spark that drives us ahead,
In every heartbeat, it's our lifeblood shed.

From the depths of the ocean to the sky above,
Oxygen sustains, a symbol of love.
It binds us together, in a delicate thread,
Uniting our spirits, as one we tread.

In every exhale, it carries away,
The burdens we bear, throughout the day.
It nourishes our cells, with each beat of the heart,
Oxygen, the force that sets us apart.
   So let us cherish this element divine,
For without it, life would cease to shine.
Oxygen, the essence of our very being,
A testament to the beauty we are seeing.

THIRTY-NINE

LIFE-GIVING GAS

Oxygen, the life-giving gas,
Without it, we would not last.
It fuels the fire in our souls,
And propels us towards our goals.

It's present in the air we breathe,
And in every plant and tree.
It's a key component of life,
Without it, there'd be only strife.

Oxygen, you are the spark,
That ignites the flame in our hearts.
You keep us alive and well,
A vital part of every cell.

So let us praise this element,
And honor its significance.
For without oxygen, we'd be lost,
And our dreams would be but dust.

ABOUT THE AUTHOR

Walter the Educator is one of the pseudonyms for Walter Anderson. Formally educated in Chemistry, Business, and Education, he is an educator, an author, a diverse entrepreneur, and the son of a disabled war veteran. "Walter the Educator" shares his time between educating and creating. He holds interests and owns several creative projects that entertain, enlighten, enhance, and educate, hoping to inspire and motivate you.

Follow, find new works, and stay up to date
with Walter the Educator™
at www.WaltertheEducator.com

www.ingramcontent.com/pod-product-compliance
Lightning Source LLC
LaVergne TN
LVHW052001060526
838201LV00059B/3765